For Emma

The
ORGANIC
CHEMISTRY
ALPHABET
BOOK

Education is the most powerful weapon
you can use to change the world."

– Nelson Mandela

A
Alkane

Alkanes contain only carbon and hydrogen atoms, and they are the simplest organic compounds.

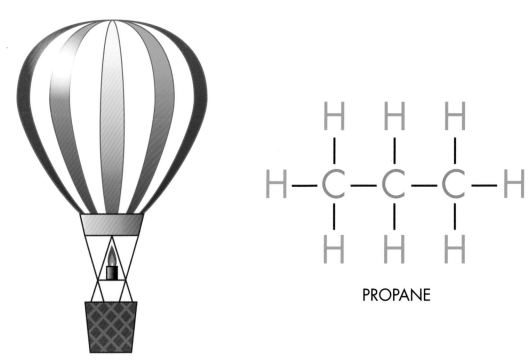

PROPANE

Organic compounds are molecules that contain carbon atoms. Alkanes have only single bonds between carbon and hydrogen atoms, with one pair of electrons shared between the carbon and hydrogen atoms. Hot air balloons rise up in the sky when the air under the balloon is heated by burning propane – an alkane that has three carbon atoms.

Benzene

Benzene has six carbon atoms that are joined in a ring and are bonded to one hydrogen atom each.

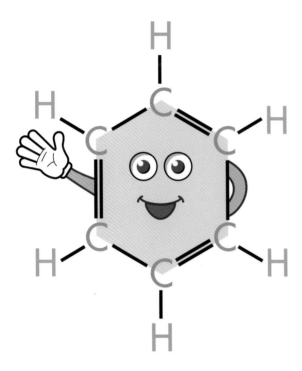

Benzene is one of the most widely used chemicals in the world. It is a compound with alternating single and double bonds between carbon atoms. The two carbon atoms in a single bond share one pair of electrons; in double bonds, the two carbon atoms share two pairs of electrons. Benzene can be formed by natural processes such as volcanic eruptions, and is an important part of gasoline.

Carbon

Carbon is a non-metallic element that is essential for all life.

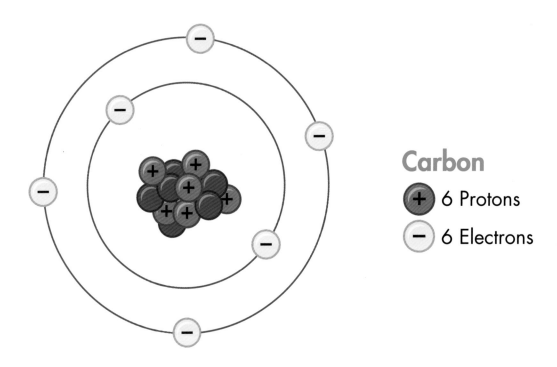

Carbon

+ 6 Protons

− 6 Electrons

Carbon has four bonding electrons – found in the outermost shell – allowing it to bond to four other atoms. Organic molecules get their diversity from the many different ways carbon can bond to other atoms. Carbon forms more compounds than any other element, with almost 10 million known compounds!

Diene

Dienes are hydocarbons that have two or more carbon-carbon double bonds.

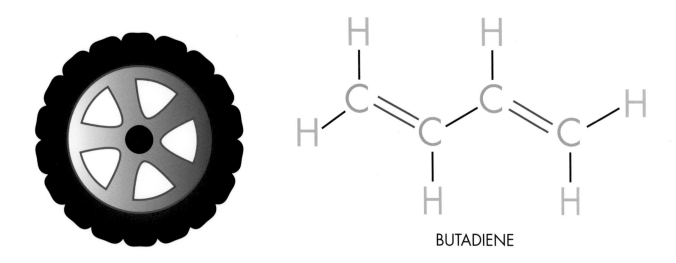

BUTADIENE

Hydrocarbons are molecules that have only carbon and hydrogen atoms. Butadiene is a molecule that has two carbon-carbon double bonds, while polybutadiene is a compound that has many butadiene molecules linked together, sometimes as many as 2,000! It is a man-made rubber that is very resistant to wear, and is used to make tires.

Enantiomer

Enantiomers are molecules that are mirror images of each other but are not the same.

Enantiomers are isomers – molecules that have the same chemical formula but different structures. Your right and left hand are mirror images of each other – they look identical when viewed in a mirror. However, when you put your left hand on top of your right hand, it is clear that they are not actually the same.

Functional Group

A molecule's functional group is the characteristic and reactive portion of that molecule.

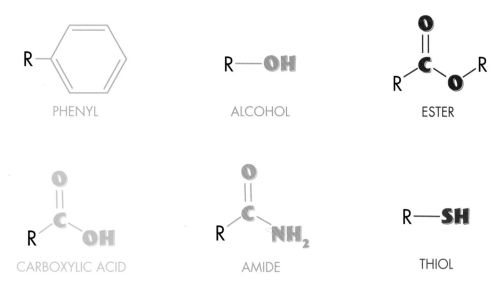

PHENYL

ALCOHOL

ESTER

CARBOXYLIC ACID

AMIDE

THIOL

Functional groups are groups of atoms that are bonded to the main carbon structure of an organic molecule, and are usually the key part that determines how the molecule will react. When functional groups are shown, the main carbon structure is shown as "R".

Grignard

Victor Grignard developed a reaction that uses magnesium (Mg) to make carbon-carbon single bonds.

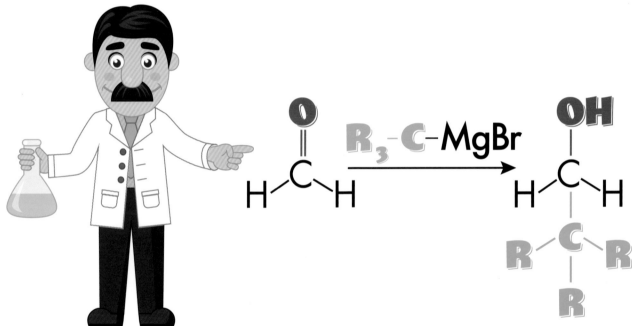

In 1912, Victor Grignard won the Nobel Prize in chemistry for developing the Grignard reaction. This is a two-step reaction where the Grignard reagent – Mg coordinated to a chlorine (Cl^-) or bromine (Br^-) ion and bonded to a carbon group – is formed in the first step. This is followed by adding a ketone or aldehyde to the reaction, resulting in a new carbon-carbon single bond. This reaction is especially important when making larger organic molecules from smaller molecules.

Hydride Anion

The hydride anion is a hydrogen atom with two electrons and a negative charge.

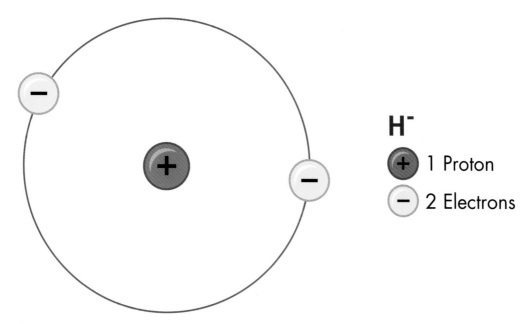

H⁻

$+$ 1 Proton

$-$ 2 Electrons

An anion is an atom with more negatively-charged electrons than positively-charged protons, and has an overall negative charge. The hydride ion is the simplest possible anion. These anions can form bonds with almost every element on the periodic table to make "hydride" molecules.

Imine

An imine is a molecule containing a double bond between a carbon and nitrogen atom.

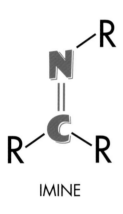

IMINE

Imines are very important for chemical reactions in the human body. One of those reactions is making a molecule called serotonin, which is found in the brain and makes people feel happy.

J-Coupling

J-coupling provides information about how atoms are connected to each other in a molecule.

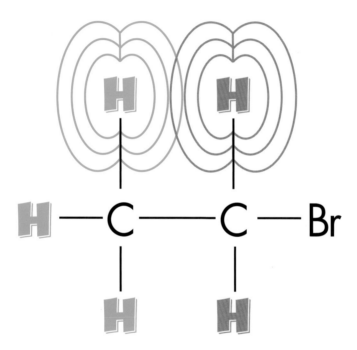

Coupling occurs when the magnetic field of a proton (a positive charge in an atom's nucleus), is influenced by the magnetic field of nearby protons. The coupling constant – J – shows how big this interaction is. The smaller the value of the J constant, the farther away the protons are from each other. This information can help chemists figure out the structure of an unknown molecule.

Ketone

Ketones are molecules that have a carbon-oxygen double bond attached to two other carbon atoms.

H_3C —C— CH_3

O

ACETONE

A ketone is another example of a functional group. A variety of ketones are made on a huge industrial scale. One of the most important ketones is acetone. Acetone is the smallest and simplest ketone, and is commonly used in nail polish remover.

Leaving Group

A leaving group is an atom or group of atoms that breaks away from the rest of the molecule during a chemical reaction.

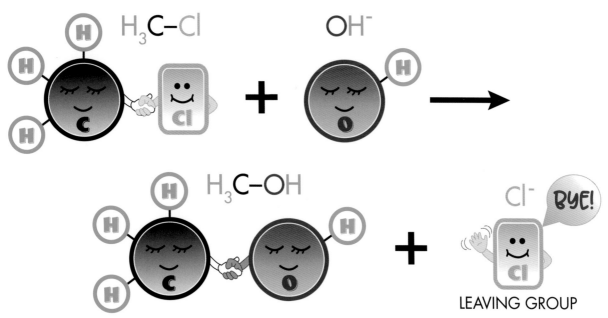

$H_3C–Cl$ + OH^- ⟶

$H_3C–OH$ + Cl^- BYE!

LEAVING GROUP

Leaving groups can be neutral, having no overall charge, or have a negative charge. Common leaving groups are the halide ions, such as chloride (Cl^-) and bromide (Br^-), found in Group 17 of the periodic table. Good leaving groups can lead to faster reactions.

Methane

Methane is the simplest organic molecule that contains only carbon and hydrogen atoms.

METHANE

Cows release a lot of methane through burping and farting – 100-150 liters a day! Scientists are trying to find ways to decrease the amount of methane produced from cows because it is one of the biggest contributors to greenhouse gas emissions.

Newman Projection

A Newman projection shows a carbon-carbon bond from "front to back".

A Newman projection is an alternate way of looking at a molecule in three dimensions (3D). To create a Newman projection, we rotate the molecule by "pulling" the front carbon in the carbon-carbon bond forward while "pushing" the back carbon backwards until the front carbon is directly in front of the back carbon – the front carbon is shown as a dot and the back carbon is shown as a circle. These projections make it easy to compare each atom's place along a bond.

Ortho Position

In a benzene molecule, when two functional groups are on neighboring carbons (C1 and C2), the groups on C1 and C2 are "ortho" to each other.

ortho-DIBROMOBENZENE
1,2-DIBROMOBENZENE

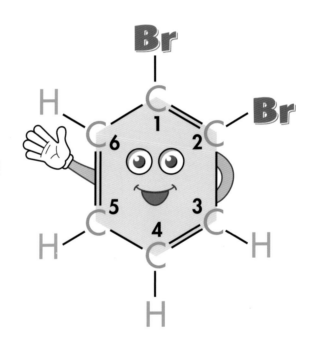

"Ortho-" is a prefix used in organic chemistry to show the positions of two functional groups relative to one another in a ringed molecule, like benzene. If the two functional groups are bonded to C1 and C3, then these groups are "meta" to each other; if the groups are bonded to C1 and C4, then they are "para" to one another. Where these these groups are positioned in a molecule can affect a chemical reaction.

Polymer

A polymer is a compound made up of many repeating units called monomers.

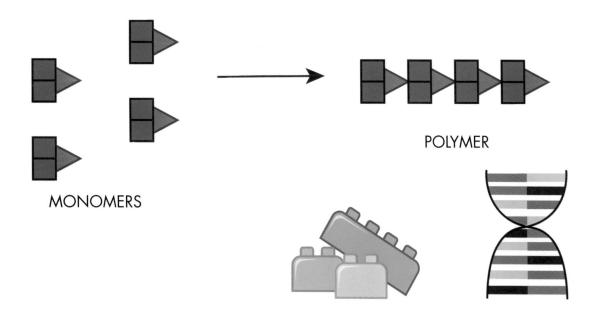

MONOMERS

POLYMER

Polymers are everywhere! Polyesters – polymers that have the ester functional group in their monomer – make up synthetic plastics like water bottles and toy blocks. Polyamides, which have monomers containing the amide functional group, are biopolymers and make up DNA and proteins found in living organisms.

Quaternary Carbon

A quaternary carbon is a carbon atom bonded to four other carbon atoms.

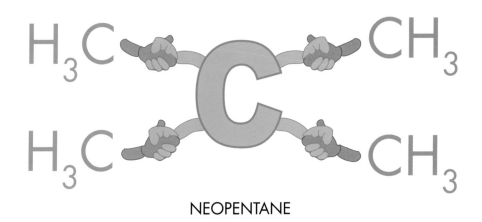

NEOPENTANE

Carbons can bond up to four atoms, and in quaternary carbons all four of those atoms are carbons. They are only found in branched organic molecules having at least 5 carbon atoms like neopentane, where the middle carbon is bonded to the carbon atoms of four methyl (CH_3) groups.

Radical

Radicals are highly reactive molecules that have an unpaired electron.

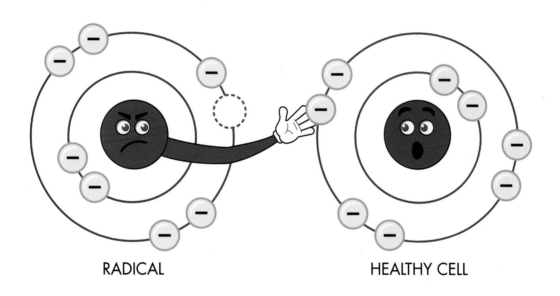

RADICAL   HEALTHY CELL

Electrons like to be in pairs. In the human body, oxygen radicals are missing an electron so they search the body trying to find other electrons in order to become a pair. This is important in many functions in the human body, but sometimes radicals can take electrons from healthy cells and harm them. Many forms of cancer are thought to be a result of radicals taking electrons from DNA, which damages the DNA molecule.

S
Saturation

When all the carbon-carbon bonds in a molecule's main structure are single bonds, it is said to be saturated.

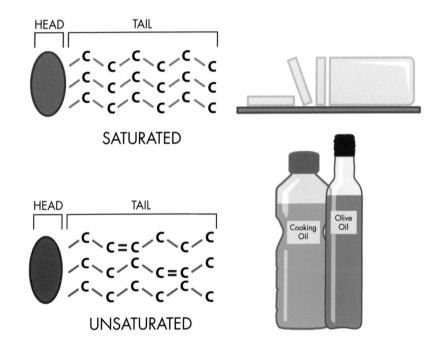

HEAD   TAIL

SATURATED

HEAD   TAIL

UNSATURATED

Cooking Oil

Olive Oil

Fat molecules are made up of a glycerol "head" bound to three fatty acid "tails". Fats are labelled as saturated or unsaturated. Saturated fats only have carbon-carbon single bonds in their fatty acid tails and are found in butter. Unsaturated fats have at least one carbon-carbon double bond in their fatty acid tails, and are found in cooking oils.

T
Thiol

Thiols are compounds that contain a sulfur atom bonded to a hydrogen atom.

Thiols are essential in living organisms. When two thiol molecules bond to each other through their sulfur atoms, they create a disulfide bond, which can form a "bridge" between two portions of a protein to help stabilize a protein's structure and function. Your hair is mainly made up of proteins called keratins, which contain numerous disulfide bonds from thiols. These bonds are part of the reason your hair is curly or straight – the "bridge" that forms creates a bend in the hair... curly hair has lots of these bridges, while straight hair has fewer.

Urethane

Urethane is used as a finish on products to protect them from weathering.

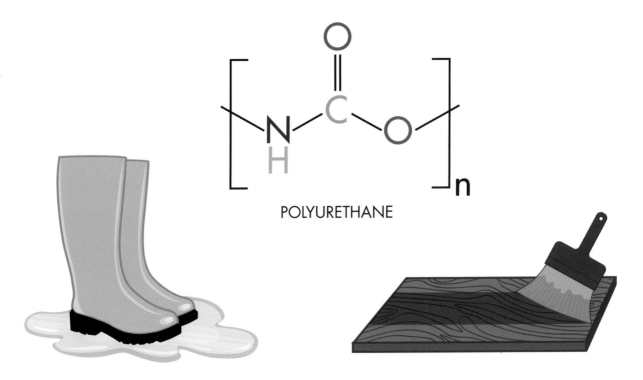

POLYURETHANE

Urethane can link together to make the common compound polyurethane ("poly" means many). The "n" shows how many urethane molecules are linked together to make polyurethane. Polyurethane is used to make surface coatings resistant against wear and tear, like in rubber boots and sealants for varnish.

Vibrations

Vibrations happen when bonds between atoms move within a molecule.

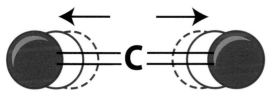

SYMMETRIC STRETCH

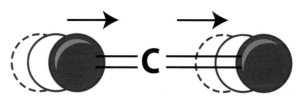

ASYMMETRIC STRETCH

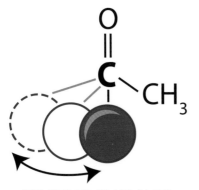

BENDING (IN-PLANE)

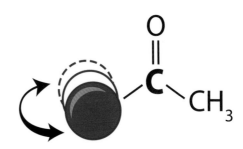

BENDING (OUT-OF-PLANE)

Molecules can have different kinds of vibrations. They can have stretching which is a change in bond length, and bending which is a change in bond angle. Each vibration has a unique energy, and therefore they can be used to help chemists determine different bonds present in a molecule.

Wittig

Georg Wittig developed a reaction to make carbon-carbon double bonds using the Wittig reagent.

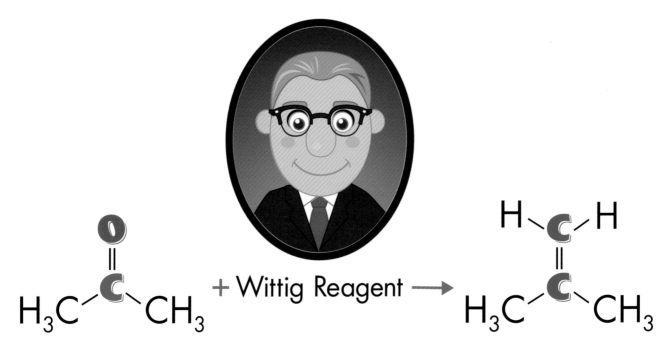

$$H_3C-\overset{\overset{\displaystyle O}{\|}}{C}-CH_3 \;+\; \text{Wittig Reagent} \;\longrightarrow\; H_3C-\overset{\overset{\displaystyle \overset{H}{\diagdown}C\overset{H}{\diagup}}{\|}}{C}-CH_3$$

In 1979, Georg Wittig won the Nobel Prize for discovering a way to make alkenes – compounds with carbon-carbon double bonds (C=C) – from molecules that have carbon-oxygen double bonds (C=O) with the help of a molecule called the Wittig reagent. The resulting molecule can then be changed further to make lots of different molecules, which makes this reaction widely used in organic chemistry.

Xylitol

Xylitol is a molecule, found in the bark of birch trees, that is used as a sweetener.

$$HO-\underset{\underset{H_2}{|}}{C}-\underset{\underset{H}{|}}{\overset{\overset{OH}{|}}{C}}-\underset{\underset{H}{|}}{\overset{\overset{|}{}}{C}}-\underset{\underset{H}{|}}{\overset{\overset{OH}{|}}{C}}-\underset{H_2}{C}-OH$$

OH

XYLITOL

GUM
SUGAR FREE

TOOTHPASTE

Xylitol is known as a sugar alcohol because of its many alcohol (OH) functional groups. It is used as an artificial sweetener in a variety of products with the most common being sugar-free gum. Xylitol is also used in toothpaste to make it taste better since xylitol does not cause tooth decay, unlike sugar.

Ylide

An ylide is a compound with two atoms bonded to each other that have opposite charges.

$$P^+ - C^- - H$$

WITTIG REAGENT

The Wittig reagent is an ylide. Ylides are commonly reagents – compounds added to a reaction to cause a chemical change, or reactive intermediates – compounds that are made and then used up during the reaction. The overall charge of an ylide is zero because the positive and negative charges cancel each other out.

Ziegler-Natta Catalyst

The Ziegler-Natta catalyst is used to make linear (straight) polymers.

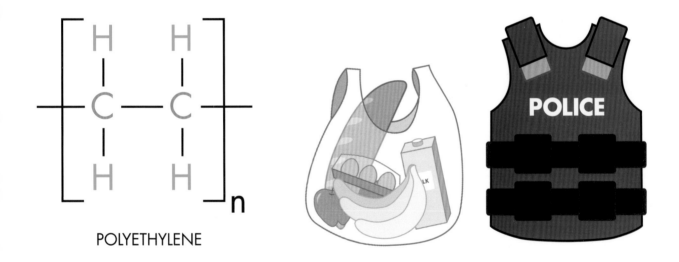

POLYETHYLENE

The Ziegler-Natta catalyst – the most common commercial catalyst – helps make the polymer polyethylene. A catalyst is a molecule that helps make a chemical reaction go faster. Polyethylene is flexible and is used to make plastic bags. High density polyethylene can be used to make very strong fibers for bulletproof vests.

Also by Christi Sperber:

The General Chemistry
Alphabet Book